How To

BUILD A TIME MACHINE

By HAZEL RICHARDSON

Illustrated by
Alan Rowe

FRANKLIN WATTS
A Division of Scholastic Inc.
New York Toronto London Auckland Sydney
Mexico City New Delhi Hong Kong
Danbury, Connecticut

For John, who was in the right place ten years too late

First published 1999 by Oxford University Press
Great Clarendon Street, Oxford OX2 6DP

First American edition 2001 by Franklin Watts
A Division of Scholastic Inc.
90 Sherman Turnpike
Danbury, CT 06816

Catalog details are available from the Library of Congress
Cataloging-in-Publication Data

ISBN 0-531-14644-8 (lib. bdg.) 0-531-13999-9 (pbk.)

Cover illustration: Andy Cooke

Printed in China

Contents

HOW TO TRAVEL THROUGH TIME

About one hundred years ago, H.G. Wells published a book called *The Time Machine.* The hero of the story tells his friends that he has discovered a way to travel through time. He has built a wonderful metal pod, with lots of dials and instruments, which he uses to travel thousands of years into the future to see what the world will be like. He finds a group of simple, fun-loving humans who live aboveground. He also finds a group of hairy people who live in underground tunnels—and they steal his time machine!

Scientists used to think time travel like this was impossible. But in 1970, astronomers made an amazing discovery. They found that pieces of a large star were being dragged off by an unseen and unbelievably powerful force. They called it a black hole. Black holes are the most terrifying things in the whole universe. They are created when enormous stars die and suck everything nearby into them. They are so powerful that not even light can escape! But black holes may be useful, too. If you go through one—and manage to avoid being squashed to death— you'll be able to travel through time.

To be a time traveler, you'll have to be very brave. It will be the most dangerous and terrifying journey you ever make—and you might not be able to get back! If you're determined to do it in spite of all the risks, this book will tell you all about

- time and how to measure it
- how to see things that happened millions of years ago
- the ways people in the past tried to see into the future
- how space and time are joined together
- how black holes are made
- how to build a time machine

CHAPTER ONE

WEIRD WAYS TO MEASURE TIME

Believe it or not, time hasn't always existed! About fifteen billion years ago, there was a huge explosion larger than anything you could ever imagine. It was bigger than all the nuclear weapons in the world going off at once. That was when time began.

The unbelievably hot clouds of dust that shot out from the Big Bang slowly turned into the stars and galaxies that make up our universe. Some scientists think the universe will shrink back again to nothing in billions of years, and then time will run backward until it stops.

Taking Your Time ....

People use time to measure how long something takes to happen. You probably don't think about it, but you use time all the time. It marks important events, such as when you are born and when you die, and everyday things, such as when you have to wake up and when you go to school.

In ancient times, it was important for people to know what time of year it was so they could successfully sow and harvest their crops. They also had to prepare for the heat and dryness of summer or the freezing cold of winter. Getting things like this right was a matter of life and death.

Ancient peoples measured time in the same way we do today—taking measurements of how the Earth moves around in space. They realized that the sun, moon, and stars seem to always move around in the sky in the same way. They noticed there was one full moon every 29½ days, and they used this to divide their year up into months of about 30 days each.

Kooky Calendars..............................

The Egyptians noticed that a bright star called the Dog Star rose in the same place next to the sun once every 365 days. They didn't know this was because the Earth was moving in a circle around the sun once a year, but it was a way of keeping regular time. They designed the first calendar in 4236 B.C. (Before Christ), the earliest recorded year in history.

Aside from the Egyptians, most people in ancient times had a calendar made up of months of 29 and 30 days. But there was one big problem. You can't fit a whole number of 30-day months into a 365-day year. (If you don't believe me, try it.) So, years based on 30-day months had only 360 days. This was great if you didn't like waiting 365 days for your birthday. But soon the years got out of rhythm with the seasons, and months that should have fallen in the summer ended up in the winter!

In desperation, some people, such as the Babylonians, added an extra day every now and then to keep the months in time with the seasons. Others, such as the ancient Greeks, added extra months.

In the seventh century B.C., Romans used a calendar that had 304 days and only 10 months. They added an extra month to the calendar every two years. To make things even more complicated, they had only three dates in each month:

- the calends (the first day of the month)
- the ides (the middle of the month)
- the nones (the ninth day before the ides)

To tell someone the date of any other day, you had to count backward from one of these dates. Sometimes the ides was on the fifteenth of the month, and sometimes it was on the thirteenth!

The people in charge of adding the extra days and months confused things even more. They sometimes added more and more months to make sure they could stay in power longer!

14

Then, in 45 B.C., Emperor Julius Caesar came along.

Caesar rejected the lunar calendar in favor of one based on the sun. He created the Julian calendar, with years that had 365 days divided into 12 months. Every 4 years, an extra day was added because a year is really 365¼ days long. When a year had an extra day, it was called a leap year. We still have leap years today.

Caesar Says...

It was Julius Caesar who gave the months the names we're familiar with today. He even named one of the new months after himself. Which one? You guessed it—July. (The next Roman emperor, Augustus, decided to get in on the act and name one of the new months after himself, too—August.) When it came to naming the days of the week, Julius named them after the Roman gods. We still use this idea, although some of the English names follow the Norse (Viking) equivalents.

Monday means "day of the moon."

Tuesday was named after me, Mars—the god of war. In English, Tuesday is named after the Norse god of war, Tiu.

16

Despite Caesar's hard work to make an accurate calendar, it was longer than the real year by 11 minutes. This doesn't sound like a problem—and it wasn't, at first—but by 1582, the dates were 10 days earlier than they should have been.

This is Pope Gregory XIII. He had a brainstorm about an easy way to solve this problem: drop ten days from the calendar! He also figured out that the turn of each millennium shouldn't be a leap year—unless it was divisible by 400. His new calendar was called the Gregorian calendar (of course), and it's the one we use today. It took a long time for the calendar to catch on in some countries, however. For instance, it wasn't used in Britain until 1752, nearly two hundred years after it was invented. By that time, Britain was so behind the Gregorian calendar that drastic action was needed to catch up. Parliament decided that the day after September 2 would become September 14! (Lots of people missed their birthdays that year.)

Timing Time . . . .

Although it was important for ancient peoples to know what time of year it was, it wasn't so important for them to know exactly what time of day it was. This was a good thing, because when clocks were first invented, they weren't very accurate.

The first clocks were made in the Middle East about five thousand years ago. The ancient Egyptians used tall, four-sided towers called obelisks to tell the time of day. An obelisk made a large shadow that moved across the ground as the sun moved across the sky during the day. For traveling businesspeople, there was also a portable shadow clock. This was a rod with another raised rod at one end, which made a shadow as the day went on.

Be a Time-Traveling Scientist
SEE HOW THE ANCIENT EGYPTIANS MEASURED TIME

WHAT YOU'LL NEED

- ○ a compass
- ○ three small pieces of straight wood
- ○ wood glue or nails

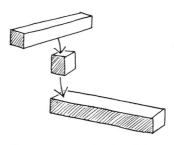

WHAT TO DO

1. Stick the pieces of wood together as shown in the picture.
2. On a sunny day, use the compass to find east and west. Line up the stem of the T-shape so that the crossbar is facing east.
3. Every hour until noon, mark the bar where the shadow of the crossbar lands on it.
4. After noon, turn the clock around so that the crossbar is facing west.
5. Check the shadow clock every hour again. Does the shadow match the marks you made earlier?

crossbar

stem

WHAT HAPPENS?

The marks you made in the morning are roughly one hour apart. You can use these marks to measure time in the afternoon, too. As long as you know which way east and west is, you will always have a portable clock!

The Egyptians thought they were very clever for coming up with their shadow clock idea, but there were two big problems with it:

1. The days get longer and shorter through the year, so clocks made during the summer were no use in the winter.

2. Shadow clocks had no way of telling the time at night or on a cloudy day.

The Egyptians had to find some other way to tell the time, so they came up with a new kind of clock that used water. If the sound of a dripping tap annoys you, you wouldn't have wanted to use one of these clocks.

The simplest dripping clocks were stone dishes with a small hole near the bottom. Water dripped out at a steady rate. As the water level went down, a mark on the inside of the dish showed what time it was. You just had to remember to fill up your clock every morning!

Be a Time-Traveling Scientist
MAKE A WATER CLOCK

WHAT YOU'LL NEED

- ○ an old plastic bottle—the larger the better
- ○ a pin
- ○ modeling clay
- ○ a large dish for the water to drip into, or a sink to put your clock in

WHAT TO DO

1. Using the pin, make a small hole in the side of the bottle, close to the bottom.
2. Plug the hole with a piece of modeling clay until you're ready to start your clock.
3. Fill the bottle with water. Make a mark at the water level.
4. Check the time and unplug the hole.
5. After half an hour, check the water level in the bottle and mark it with a pen.
6. Plug the hole in the bottle while you measure how far the water level dropped in that half hour. Then make more marks the same distance apart to measure off more half hours.
7. Start the clock again and check it every hour or half hour. How accurate is the clock?

Clocking On..................................

Even though water clocks are a bit messy, they're quite accurate. They were the most popular way of measuring time for thousands of years. The great scientist Archimedes even had a water-powered alarm clock.

Time to get up!

Over the centuries, people came up with other ideas for measuring time, such as huge egg timers or candles with marks on them that showed how long they had been burning. But none of these clocks could tell you *exactly* what time it was. Then a man named Galileo Galilei made an amazing discovery...

Swinging Seconds
Italy, 1582

Here's Galileo at a Sunday service in the cathedral. He's bored stiff and begins to daydream. He looks up at the ceiling and sees that one of the lanterns is swinging back and forth. As he watches, he realizes something.

It always takes the same time for the lantern to swing back and forth!

Excited, Galileo runs home and experiments with weights on long pieces of string. He's right—the weight always takes the same time to swing. He has invented the pendulum! He thinks this might be a good way to measure time.

Unfortunately, Galileo doesn't get to use his pendulum to make a clock because he gets more interested in space—a big mistake! He invents a very powerful telescope and decides that the Earth moves around the sun. This idea was already suggested in 1530 by an astronomer named Copernicus, who was ridiculed. Now Galileo is getting laughed at, too.

But Galileo knows he is right and writes a book about his new ideas. This gets him into a lot of trouble. Italy is very religious, and people think the sun must go around the Earth because God made the Earth especially for us. Galileo is called a troublemaker and warned not to say anything else about the Earth going around the sun—or else!

Galileo manages to keep his mouth shut for years. Then, when he thinks the commotion has died down, he writes another book. The book becomes a best-seller, and he is arrested by the Inquisition! He is put on trial for daring to say that the sun is the center of the solar system, and he is found guilty. Galileo is kept under house arrest for the rest of his life.

Because of this, the first accurate clocks using pendulums are not invented until the 1600s.

Be a Time-Traveling Scientist
USE A PENDULUM TO MEASURE TIME

WHAT YOU'LL NEED
- ⭘ a digital watch or stopwatch
- ⭘ a long piece of string or a shoelace
- ⭘ a weight, such as a lump of clay

WHAT TO DO
1. Tie the weight onto the piece of string, and tie the other end to a beam or a hook.
2. Pull the pendulum back and let it go. Measure the time the pendulum takes to swing forward and backward ten times.
3. Stop the pendulum and start it swinging again. Measure the time it takes to make ten swings.

WHAT HAPPENS?
As long as the swings are fairly small, the pendulum always takes the same time to make ten swings. The only way you can change this is to make the string longer or shorter. (Try it and see.)

Pendulum Power ·······························

In 1656, the first accurate clocks were invented by the Dutch scientist Christian Huygens, who used a motor to keep the pendulum moving. His first clocks could tell the time to within a minute a day! He made improvements and soon had clocks that were accurate to within 10 seconds a day. Although pendulum clocks were useful for timekeeping, they had one very big drawback—they were way too large to carry around.

Have you got the time?

Luckily, two years later, Robert Hooke found a way of powering clocks so they could be made much smaller and even worn as a watch. This was a spring clock, which we still use today. Inside the clock or watch is a small, round spring that slowly unwinds. This moves a wheel in the clock that connects with other wheels to turn the hands. Brilliant! (But useless if you forget to wind it up.) These were the most accurate clocks until the 1930s, when the quartz clock was invented.

Quartz is a crystal that does something really strange. When an electric current hits it, it changes shape and gives off an electric charge of its own.

A quartz clock works because the quartz crystal inside it is part of an electric circuit. As the electric current flows through the crystal, it vibrates and gives off an electric signal to operate the clock's display.

Quartz clocks are very accurate, but not nearly as accurate at measuring time as the most amazing clocks we have today—atomic caesium clocks. Atomic caesium clocks are correct to about one-millionth of a second a year!

Caesium is a chemical element. The atoms (building blocks) in it absorb any radiation that hits them and give some off. This taking in and giving off of radiation is called resonance. Every caesium atom in the entire universe resonates at the same rate, and this never changes. In 1 second, a caesium atom resonates exactly 9,192,631,770 times!

In 1967, the world began to use caesium as the official international timekeeper, so all the clocks in the world are checked against an atomic caesium clock. It's a good thing we did this, because the caesium clock actually tells the time better than measuring it by the Earth spinning around! This is because the time it takes the Earth to spin around every day is always slightly different. You won't notice it, but every day is always a few thousandths of a second longer or shorter than the one before it.

Because the caesium atoms and the Earth sometimes differ in their measurement of time, leap seconds are added. If you listen to the radio, you can very occasionally hear the leap second being added on. There are beeps before the news, and at special times an extra beep is put in. This is a leap second.

CHAPTER TWO

THE PAST AND THE FUTURE

Time is like a long, one-way road stretching out in front of you as far as you can see. As your life goes on, you move along the road in one direction. You can remember where you've been, but you can never go back there. You can imagine where you might be going in the future, but you can't see what it is really like until you get there. And however much you would like to, you can't get off the road and stop for a while.

Although you can't go backward along the road of time, there are ways to see things actually happening in the past...

Go outside and look at the sun. Did you know you are looking at something that happened 8 minutes ago? The sun is 92 million miles (149 million kilometers) away from us, and it takes the sun's light 8 minutes to reach Earth. If the sun suddenly went out, we wouldn't know about it until 8 minutes after it had happened!

You can! The universe is so enormous that it takes millions of years for the light from some stars to reach us on Earth. And the farther away the stars are from

Earth, the longer it takes for their light to reach us. In fact, some of the stars you see each night might no longer exist!

Twinkle, twinkle, little... nothing!

If you could travel faster than the speed of light, you would be able to zoom away from Earth and see things that happened in the past. For example, imagine that you want to see the signing of the Declaration of Independence in 1776. The light that bounced off Jefferson's pen that day is now 225 years away from Earth. So, all you have to do is hop in your spaceship and zoom to a planet about 225 light years away. (A light year is the distance light travels in one year.) With a very powerful telescope, you could watch every signature! You could travel even farther and watch the pyramids being built, or see if the dinosaurs really *were* made extinct by an asteroid.

Unfortunately, there are two problems with this idea:

1. Scientists think it's impossible to travel faster than the speed of light—no matter what you see in sci-fi movies.

2. Even if you could go back in time, you wouldn't be able to get involved in what was happening, because you'd be miles away in space. (This is why you'd need a very, very powerful telescope to watch all the details of Independence Day.) Therefore, it wouldn't be possible to watch the lottery results on a Saturday and go back to the Friday before so you could win, because you'd be too far away to buy a ticket.

Seeing the Future ....

For thousands of years, people have dreamed about
seeing into the future. Most people want to know
what is going to happen to them. If you could travel
through time, people would beg you to come back
and tell them what was going to happen. There have
always been people who claim they don't need a time-
travel machine to see the future. These people are
called fortune-tellers, soothsayers, or clairvoyants.
They all use different ways of seeing into the future,
such as crystal balls, tarot cards, or astrology. Some
fortune-telling methods are especially strange:

Throwing a bunch of sticks into the air and interpreting which way they fall (as done by the ancient Chinese)

Looking at the steaming intestines of animals

Heating tortoise shells (without the tortoise inside!) to see what pattern they crack into. If you can't persuade a tortoise to lend you its shell, the shoulder blade from a cow will do (although the cow might not be too happy either).

Watching how birds fly

Reading the pattern of tea leaves in the bottom of a tea cup

What a yummy cup of tea!

I didn't know she switched to tea bags!

Be a Time-Traveling Scientist
SEE HOW FORTUNE-TELLING CAN FOOL PEOPLE

Some people take fortune-tellers seriously because they get some things right by chance. People generally remember what fortune-tellers get right, but rarely what they get wrong. You can prove this in a very simple experiment.

WHAT YOU'LL NEED
- ⭕ a pack of playing cards
- ⭕ some volunteers
- ⭕ a notebook
- ⭕ a silly outfit and a wacky name, such as Bonzo the Great

WHAT TO DO
1. Get your volunteers to ask you about something that might happen to them in the next week or so, for example:
   "Will I win some money?"
   "Will I get a new girlfriend?"
   "Will I pass the test at school?"
   "Will my dad let me go to the concert?"
   Write down the questions in a list along with the name of the person who asked each one.
2. Shuffle a pack of cards really well. For each question, a card out of the pack.
   If it is a black suit, then write the answer as "no."
   If it is a red suit, write the answer as "yes."
3. Tell your volunteers your predictions.

4. In a couple of weeks, ask them what really happened. Compare what happened with what you predicted. How do your friends react?

WHAT HAPPENS?
You should find that you did predict what would happen to some of your friends—by chance—and they are amazed.

THE GREAT GUIDO IS UNAVAILABLE TODAY DUE TO UNFORSEEN CIRCUMSTANCES!

Strange Signs

Fortune-tellers have come up with some weird predictions over the years!

Calchas was a very famous Greek fortune-teller at the time of the Trojan War. At one point during the war, the Greek navy could not set sail because there was not enough wind. Calchas told the Greek king Agamemnon that the gods were annoyed with him. The only way to please them so they would make some wind was to sacrifice his daughter! Calchas was so highly respected that Agamemnon went ahead and killed her.

But Calchas did prove useful when he told the Greeks how they could win the war. He told the Greek army to build a huge wooden horse and hide inside it. The curious Trojans opened the gates and brought the horse into their city—just as Calchas had said they would—with all the soldiers inside!

In the twelfth century, St. Malachi made some very famous predictions. He predicted the next 111 popes, starting in 1143. (Unfortunately, none of them turned out to be correct!) St. Malachi also predicted that after the last pope, the end of the world would come. He calculated that this would be sometime around 2012—so, according to him, we don't have much more time left!

Perhaps the most famous fortune-teller of all is Nostradamus, who skyrocketed to fame in 1555 for writing a book called *Centuries. Centuries* is full of prophecies written in funny rhymes—but many of them are so vague that people can interpret them almost any way they want. For instance, some people say the rhyme below was a prediction of Hitler. (They guessed that Hitler must be "this thin man" because he was a vegetarian!)

> *Nine years this thin man will hold in peace*
> *Then he will fall into thirst so bloody*
> *For their peoples he will die without faith or law*
> *Killed by one much more mild and good-natured.*

Nostradamus predicted that the world would end in 3797—so there's still a while to go before we find out if *he* was right!

Changing History ....

People just can't wait to find out what happens in the future. Most scientists think fortune-tellers are fakes, or that they make wild guesses, but some people really believe them.

The only way to see the future for certain is to travel through time. If you could do that, you'd also be able to go back in time and actually change the past. When you begin to imagine what kind of effects that could have, it can be mind-boggling!

CHAPTER THREE

TIME TRAVEL: A RISKY BUSINESS

Many people dream about traveling to a different time, staying there for a while, and then going back to the time they left behind. This can be more dangerous than you might think. First of all, you might experience something called a paradox.

Problematic Paradoxes......................

A paradox is something that cannot happen or cannot be true. For instance, imagine that someone you meet says to you, "I always lie." Are they telling the truth?

    A. yes
    B. no

Neither answer is correct! They can't be telling the truth, because if they were they would be lying! They can't be lying, because if they were they would be telling the truth! (Think about it...)

Paradoxes can happen a lot in time travel. For instance, imagine that you've just been given a new bike for your birthday. One day, you forget to lock it when you run into a store, and it gets stolen. You're upset—but then you have a bright idea. Why don't you go back in time and tell yourself to lock the bike up? Then it won't get stolen. What a brilliant idea!

So, you go back in time, meet yourself just before you go back into the store, and say:

You do, and you think you're very clever because your bike isn't stolen.

But there's a problem here. If your bike isn't stolen, you won't need to go back in time to tell yourself to lock it up. So you don't. So your bike does get stolen. So you go back. So your bike isn't stolen. So you don't go back. And so on, forever and ever!

A more dangerous paradox is called the Grandmother-Killing Paradox. (I'll bet you can't guess what this one is about!)

It'd better not be about what I think it's about!

You get in your time machine and go back fifty years or more, to a time before your parents were born. But there's a terrible accident! Your time machine lands right on top of your grandmother while she's out shopping, and she dies.

Oops!

With Grandma dead, your mother or father will never be born. If they aren't born, *you* will never exist! Now, here's the paradox: If you don't exist, how could you have killed your grandmother on a trip from the future? You couldn't have! You have to be very careful when traveling back in time to avoid making anything like this happen.

Time Travel Trickery ....

Scientists have suggested two things that might allow you to avoid paradoxes when you travel through time. Unfortunately, none of the scientists have been time-traveling, so they don't know if either idea is right!

1. Some scientists think that when we're able to travel in time, we might find that we can't actually change anything in the past. They think we won't be able to alter anything that has already happened—including a trip back in time. So even if you go back to tell yourself to lock up your bike, something will happen to make sure that the bike is stolen. And if you did land on your grandmother in your time machine, she wouldn't be killed, but might just end up in hospital for a while. This is because you have been born at some point in time, and nothing can change that.

2. Other scientists believe in something called the Many Universes Theory. To understand what this is, we need to meet a scientist named Ernest Schrodinger.

Dead or Alive?
Switzerland, 1926

This is Ernest Schrodinger. He's famous for an experiment that involves locking a cat in a box with a bottle of poison.

Before you start getting worried about the cat, he doesn't really carry out the experiment—he just thinks about it!

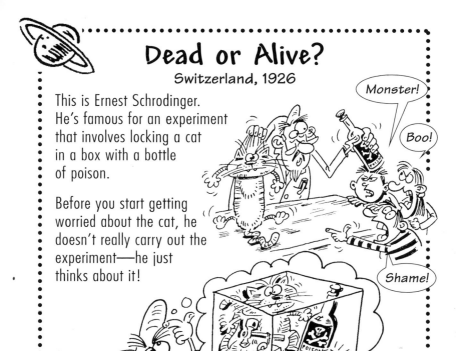

Monster!

Boo!

Shame!

Imagine if I put a cat in a box with a bottle of poison and something radioactive, like a lump of uranium. The uranium gives off radioactivity every so often, but I don't know exactly when. I also put a Geiger counter in the box. This measures radioactivity. As soon as the uranium gives off some radioactivity, the Geiger counter will measure it. When it does, it will cause a hammer to smash open the bottle of poison, and the cat will die. If I shut the cat in the box and wait until there is a 50/50 chance that the uranium has given off some radioactivity, is the cat alive or dead before I open the box?

What do you think? Is the cat:
  A. alive
  B. dead
  C. digging its way out

The cat could be alive or it could be dead. We have no way of knowing until we open the box. But Schrodinger said that things such as the radioactivity coming out of uranium are only real when we see them. Before we look in the box, it is possible that some radioactivity has been given off, but it isn't real. It is also possible that no radioactivity has been given off, but that isn't real either. One of them becomes real when we open the box and see what has happened. Before we do, both possibilities exist. Before we look in the box, there is a dead cat and a live cat in the box at the same time.

That's impossible! This Schrodinger was a nut! How can you have a dead cat and a live cat at the same time? We only put one cat in the box!

All of this does sound silly, but it is true. (And if you want to know why, you will have to study physics in college...) But how does having a dead cat and a live cat at the same time solve the problem of going back in time and squashing your grandmother? Well, you can only have the dead cat and the live cat at the same time just before you open the box. Once you open the box, only one possibility becomes real. So what happens to the other one? Well, some scientists think that once you open the box, the universe splits into two. In one universe, you open the box and the cat is alive (phew!). In the other universe, you open the box and the cat is dead.

Don't worry, Mom. In another universe, Muffy's still alive.

You are also alive in two universes—even though you don't know it.

This is the Many Universes Theory. For everything that could happen in more than one way, the universe splits up. So there are millions and millions of universes, different in tiny ways. For instance, in one universe, the dinosaurs didn't become extinct, and humans never evolved past little mouse-like animals. In another universe, Germany won World War II. In another, England beat Ireland in the 1998 World Cup soccer championship.

So, if you go back and land on your grandmother, there wouldn't be a paradox. You might kill her or you might not. If you don't, that's fine. If you do, the universe splits up and you go into a universe where you were never born. Because you had been born in another universe, you would still exist. So, the problem is solved—except for the fact that you could never get back to the one you started from!

Out of the Paradox, into the Black Hole...

Even though you might be able to get around the paradox problem, there are more dangers in store for you when you travel through time.

Imagine that you've decided to go back in time to win the lottery. You buy your ticket and wait for the numbers that you've already seen picked for the next day. You imagine all the wonderful things you are going to buy when you claim your prize.

Not so fast! You've forgotten the rules of time travel. Because you didn't buy a lottery ticket in the universe you started off in, as soon as you go back and buy one, you slip into another universe. In that universe, the numbers are different, so you haven't won.

To make things worse, there might be another you in the new universe, and you have nowhere to live and no money! It would be almost impossible to get back into the universe you started off in.

The most risky thing of all about time travel is that the best way to build a time-travel machine involves using the most deadly and powerful thing in the universe—a black hole.

CHAPTER FOUR

WHAT ARE BLACK HOLES?

Black holes are the strangest and most terrifying things in the universe. We know they are out there, but we can't see them because they are blacker than the darkest place you have ever been. Amazingly, they start off as stars brighter than the sun.

The Birth of a Star

Space, a few billion years ago

Here we are, surrounded by floating clouds of hot gas left over from the Big Bang. The gas is made up of hydrogen, and the atoms in it bump into each other and stick together. This makes a spinning ball of gas, which gets bigger and bigger and heavier and heavier. The atoms in the center of the ball get squeezed together, and the pressure makes them heat up.

The center of the ball of gas eventually reaches more than 59 million°F (15 million°C). When it gets this hot, the hydrogen atoms have enough energy to join together and turn into helium atoms. This is called nuclear fusion, and it releases an enormous amount of energy as heat and light. A star is born!

The Birth of a Red Giant.....................

Lots of new stars are enormous, with enough fuel to burn for millions of years, even though they burn millions of tons of fuel every second. (In smaller stars like our sun, the fuel can last for billions of years!) If we wait for a few million years and then return to the star we watched being born, we can see what happens when it starts to run out of fuel.

Most of the hydrogen in the star has been turned into helium, so it begins to fuse together helium atoms instead of hydrogen. This makes heavier atoms such as carbon and iron. The star goes on burning happily for a few million more years, but eventually even the helium runs out. Now the star is in big trouble, because the force of gravity begins to take over.

Gravity is the pulling force that keeps us on the ground. The heavier something is, the greater the force of gravity it has. (For instance, the moon is a lot smaller and lighter than the Earth, so it has one-sixth the gravity of Earth.) As the star burns helium and makes heavier elements, its center gets heavier and heavier and the gravity inside gets stronger.

The gravity squeezes the center of the star so hard that it gets even hotter. The outside of the star billows out to an enormous size, and the star becomes what is called a red giant.

Smaller stars, such as our sun, turn into red giants and then shrink into tiny, cooler stars called white dwarves. (This will happen to our sun in about five billion years. It will eventually end up about the same size as Earth.)

If a really massive star becomes a red giant, something even more spectacular happens. The star has burned up all its fuel and is left with only iron, which can't be used for nuclear fusion. To get the energy to fuse the iron together, the star tries to squeeze its center more—and the result is disaster! Within a few seconds, the temperature inside the star shoots up to the unbelievable level of 122 billion°F (50 billion°C) and the star is ripped apart. This explosion, as bright as a billion suns, is called a supernova.

Astronomers watched a star explode like this about three hundred years ago. The cloud of dust it threw off is called Cassiopeia A, which still shines today.

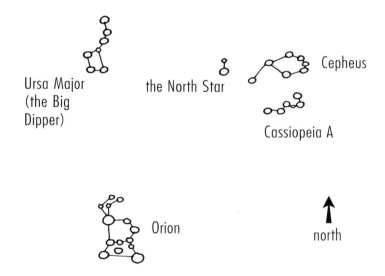

Ursa Major
(the Big
Dipper)

the North Star

Cepheus

Cassiopeia A

Orion

north

When a massive star is ripped apart like this, the core of the star is left behind. Gravity within it is so strong that not even light can escape from its surface. Nothing can prevent the core from collapsing in under its own gravity, and it turns into a black hole.

Black holes are usually made when a star weighs more than three times as much as the sun. Can you imagine how gigantic these stars are? A star that is 1.5 million miles (2 million km) across can be squeezed to only a couple of miles across! All the mass in the center being squeezed into such a small space makes the force of gravity enormous.

Be a Time-Traveling Scientist
SEE HOW BEING SQUEEZED CAN MAKE A STAR EXPLODE

WARNING
This is a REALLY messy experiment, so ask permission first.

WHAT YOU'LL NEED
- ○ an orange (this is your pretend star)
- ○ a clamp
- ○ something to clamp your orange to

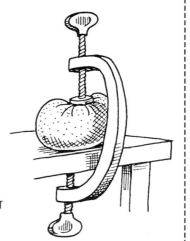

WHAT TO DO
1. Clamp the orange to a table or other surface.
2. Slowly turn the clamp tighter and tighter and tighter.
3. Watch what happens.

WHAT HAPPENS?
The orange explodes. Happy cleaning!

The Discovery of Black Holes ....

If the gravity in a black hole is so strong that not even light can escape, how do we know they exist?

Scientists first predicted the existence of black holes over two hundred years ago. In 1784, John Michell guessed that gravity had an effect on light. He said that if a star were big enough, it would have such a strong gravitational pull that not even light could escape from it. People thought this was a weird idea.

Then, in the twentieth century, astronomers discovered neutron stars. A neutron star is sometimes made when a star blows up in a supernova explosion. The star might have started off as large as our sun, but ends up only a few miles across. Neutron stars are so dense that one as big as a pea would weigh more than 100,000 tons!

After the discovery of neutron stars, scientists became very unhappy. They did not really want to believe that they had been wrong all the time and that black holes could actually exist. A neutron star would only have to shrink to a third of its size, and it would be dense enough to be a black hole. Could enormous stars really collapse like this?

Then a brilliant scientist, Albert Einstein, came along.

Einstein Begins to Bend the Rules

Germany, 1916

Albert Einstein was one of the greatest scientists of all time. When he was in school, his teachers thought he was moody and untalented. Even so, Einstein told them that one day he would solve all the riddles of the world.

He didn't do very well in college, either, and his first job was working as a clerk in a patent office. But in his spare time he worked on physics problems. In 1916, Einstein published a groundbreaking idea called the Theory of Relativity, in which he stated that space and time could be bent by gravity.

A month after Einstein told the world about his amazing theory, a German scientist named Karl Schwarzchild made an amazing announcement. He guessed that space could be bent so much that a part of it could be cut off from the rest of the universe. Schwarzchild had discovered that black holes could exist. Something really heavy could bend space so much that anything falling into it would never be able to get out.

Be a Time-Traveling Scientist
SEE HOW GRAVITY CAN
BEND SPACE

WHAT YOU'LL NEED
- ○ a pair of nylon pantyhose with one leg cut lengthwise so it can be opened up
- ○ a table
- ○ a tennis ball
- ○ strong tape
- ○ some marbles

WHAT TO DO
1. Turn the table upside down. Stretch the opened-out pantyhose leg over the table legs and tape it tightly to the four legs so that the pantyhose lies flat. This is pretend space. (Note: you'll use this setup in later experiments.)
2. Roll some marbles across the sheet and see how they roll.
3. Now put the tennis ball (your pretend sun) in the center of the pantyhose and roll the marbles across it again.

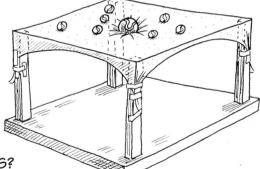

WHAT HAPPENS?
Space is usually flat, like the pantyhose stretched over the table. But when something heavy like the tennis ball (the pretend sun) is in it, space is bent. Anything going past it moves in a different way.

Be a Time-Traveling Scientist
MAKE A BLACK HOLE

To see how a black hole works, you'll need to use the pretend space model (the pantyhose on the table) that you used in the last experiment.

WHAT YOU'LL NEED
- ○ the space table
- ○ a heavy ball bearing
- ○ marbles
- ○ a tennis ball

WHAT TO DO
1. Put the tennis ball in the center of the pantyhose and roll the marbles past it. What happens to them?
2. Now put the ball bearing in the center of the pantyhose instead. This is a pretend black hole. Roll the marbles across the pantyhose at different distances from the black hole. What happens to them?

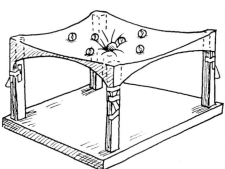

WHAT HAPPENS?
When space is bent just a little, the marbles change direction slightly, but still manage to run past the pretend star. When space is bent more strongly by the black hole, what happens to the marbles depends on how close they pass by it. If they are far enough away, they manage to get past it. If they get a bit closer, they are pulled toward it and zoom around it in a circle. If they get too close, they are sucked in and cannot get out again. This is exactly what happens to light when it gets too close to a black hole.

Seeing the Invisible............................

Even though scientists knew black holes could exist, the first one wasn't found until 1970, when a special satellite called Uhuru was launched into orbit. The job of this satellite was to find anything in the universe that was giving off X rays. X rays are far more powerful than light rays (which is why they can go through your body!) but can't get through the Earth's atmosphere. Once Uhuru was in space, it came across a strange object that had so much energy that it was giving off very powerful X rays. At the same place, scientists could see a huge star, about thirty times as large as the sun. This star was being dragged around in a circle by something that couldn't be seen, but that must have weighed ten times as much as the sun.

Astronomers realized they had found their first black hole. They called it Cygnus X-1 because it was the first X-ray source found in the star constellation Cygnus. It is about six thousand light years away from us. This black hole is about 19 miles (30 km) across and is so close to another star that it is pulling a steady stream of gas from this neighboring star into itself. As the gas falls into the black hole, it speeds up to almost the speed of light and gets very hot. Just before it falls into the black hole, it is so hot that it gives off the X rays that show us a black hole exists.

Some black holes are even closer to us than Cygnus X-1 is. In 1989, British astronomers discovered that there is even one in our own galaxy, only about fifteen light years away from us. A star smaller than our sun is being dragged around an unseen object weighing more than twelve times as much. This black hole is called V404 Cygni.

The Weight of a Black Hole.....................

Even though we can't see black holes, if they are close to a star we can figure out how much they weigh from the force of their gravity. The more gravity they have, the faster the nearby star is whirled around them. The black holes we have discovered are all close to another star. There must be many more black holes throughout the universe, but we can't find them yet because they are not close enough to another star for us to detect them. These are silent and deadly black holes, lying in wait to trap any unwary space travelers who get too close to them...

CHAPTER FIVE

FALLING THROUGH TIME

What would happen to you if you fell into a black hole? Could you get out the other side? Would you be crushed to death? Could you use it to travel through time? Let's see what happens when the intrepid crew of the starship *Gorgonzola* stumbles into the path of a black hole…

Captain's Log:
The Starship Gorgonzola
Stardate 2316

After saving the inhabitants of the planet Zorgon from the dreaded three-headed Freedo monster, we have set off into a region of space that has not yet been charted.

The crew is getting nervous. This area of space seems too dark and empty. Our computer is plotting a course for the nearest star system, which is about twenty-five light years away. I am not sure why this area seems so empty, so I have sent a small scout ship ahead of us to send back information, just in case there are any dangers that the computer hasn't detected. I had to handpick a crew of five robots for the scout ship, because none of my cowardly crew were eager to risk their lives. We are following close behind *Scout Ship 5*—though not so close that we can't get out of any problems they come across!

I have just been woken up by an urgent transmission. *Scout Ship 5* is in an emergency situation! The sensors indicate that the ship is approaching an area of incredible gravitational pull—they have stumbled onto a black hole! I have ordered the scout ship to return to the *Gorgonzola* immediately and told the computer to plot a course out of the area.

From a second urgent transmission we know that *Scout Ship 5* is in trouble. The thrusters aren't strong enough to boost it away from the incredible gravity force of the black hole. The scout ship is being dragged in! The computer has calculated that the black hole is about 6 miles (10 km) across. With a black hole this dense, there is nothing we can do except wait at a safe distance and watch as *Scout Ship 5* is torn apart. We are analyzing the information sent back from the scout ship... We can see it speeding up as it approaches the black hole... The clock on *Scout Ship 5* is running at the same time as ours, but the computer estimates that their time will slow down as they get closer to the black hole because the gravity bends time as well as space...

Something strange is happening. *Scout Ship 5* is getting thinner and longer. It looks like it is turning red, and the clock onboard is running much slower than ours. The pull of the black hole's gravity is stretching the ship like a piece of taffy. The computer tells us that it's nearly forty times its original length! The clock onboard the ship has stopped at exactly 3:30...

We have lost all contact with *Scout Ship 5*. It is also very difficult to see it now. The computer says this is because the light can't escape from the gravity force. We have been trying to keep an eye on the ship for three days now, and it hasn't moved from its last position when the clocks onboard stopped. The computer says this is because time runs so slowly just outside the black hole that we could wait forever and would never see the ship fall in. So we are moving on—after leaving a warning beacon for any future starship travelers that pass this way...

As you can see, falling into a black hole is not a very pleasant experience! *Scout Ship 5* was unlucky. The smaller the black hole, the more deadly it is. The black hole that *Scout Ship 5* stumbled across was only 6 miles (10 km) wide, so it bent space very strongly. If humans had been sucked toward it, rather than robots, they would have felt as if they were hanging from the top of the Empire State Building with a thousand elephants clinging to their ankles. The force of gravity near the hole would have stretched them so much that they would have been torn apart long before falling into the hole itself.

How to Use a Black Hole ....

So, if traveling through a black hole is so dangerous, how could one be used for time travel?

The trick is to find a really enormous black hole. If you were lucky enough to come across a black hole that weighed more than ten million times as much as the sun, the taffy-stretching forces you would feel as you fell in would not be as strong as the ones *Scout Ship 5* was subjected to when it fell into its black hole. The larger a black hole is, the less steeply it bends space around it. So if you fell into a larger black hole, you would still be stretched and feel very uncomfortable, but you would not be ripped apart. (Phew!)

Be a Time-Traveling Scientist
SEE HOW YOU WOULDN'T BE TORN APART IN A LARGE BLACK HOLE

It might seem strange that it is easier to survive falling into a huge black hole than a small one. This experiment will show you why.

WHAT YOU'LL NEED
- ○ the space table you made earlier
- ○ one very large and two small marbles or ball bearings
- ○ seven rubber bands
- ○ a piece of string

WHAT TO DO
1. Tie the rubber bands together, as in the picture below, so that they look like legs on a body.
2. Knot the bottom of the legs around one of the small marbles.
3. Tie the rubber-band astronaut to the edge of the space table.
4. Put the small marble in the center of the space pantyhose. This is a pretend small black hole.

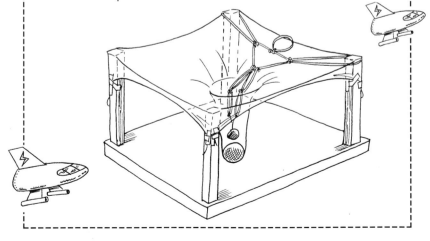

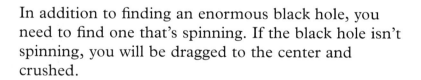

5. Push the marble attached to the rubber band legs toward the black hole and see what would happen to your legs if you fell into this black hole.
6. Now use the large ball or marble as a pretend large black hole and repeat the experiment. What would happen to you if you fell into this black hole?

WHAT HAPPENS?
The legs on your pretend astronaut are stretched less as it falls into the large black hole because the dip into the black hole is not as steep. This is why you need to choose your black hole carefully!

In addition to finding an enormous black hole, you need to find one that's spinning. If the black hole isn't spinning, you will be dragged to the center and crushed.

So, you've found the perfect black hole, and you're ready to head inside it? Let's go!

CHAPTER SIX

FROM TIME TO TIME: BRIDGING THE GAP

So, you're falling into an enormous black hole as the crew on your spaceship watches from a safe distance. Even though they see you get closer and closer to the edge of the hole, they will never see you fall in. Time slows down so much around the black hole that the ship seems to stop just before entering it. However, you don't notice this. To you, time seems normal.

Now you will need all your skill to avoid the very center of the black hole, where you would be crushed to death instantaneously. If you manage to do this, you might be lucky and find a special pathway called the Einstein-Rosen bridge.

When Albert Einstein thought up the theory of relativity, he didn't believe black holes could exist. So when other scientists proved their existence, he became a little upset. He decided he was going to be the first to find out all about black holes. He worked with another scientist named Nathan Rosen and discovered something amazing: a black hole can open out at the bottom to join up with another part of space—or even a completely different universe! This passage through the black hole is called an Einstein-Rosen bridge or a wormhole.

Before you start panicking, scientists don't believe there are giant worms lurking in the middle of bottomless black holes waiting to munch on intrepid space travelers. (Then again, since they've never been through one, how would they know?) So don't worry about *that* danger. Concentrate on the many others instead!

Wiggling Through a Wormhole ....

The idea of using wormholes as a way of traveling through time was thought up by American scientist Kip Thorne in 1985. Thorne also figured out that a wormhole should allow two-way travel, so if you don't like the time you end up in, you can still get back again.

Before you go diving into the wormhole in your spaceship, there's something you should know. Wormholes are very delicate and don't like people going through them. When they see a spaceship coming, they close up! As the spaceship falls into the black hole, it is pulled faster and faster by the force of gravity. It speeds up, and this makes the spaceship give off energy waves. These energy waves move faster than the spaceship and reach the wormhole before the spaceship. Then disaster strikes! The energy waves disturb the delicate wormhole and it slams shut, leaving you and your spaceship to crash into the center of the black hole. Ouch!

Did I say it was impossible to travel through a wormhole? No, I didn't! It *is* possible, but you have to find a way to keep the wormhole open so that you can sneak through. To do this, you need something that will cancel out the waves your spaceship makes as you move through the black hole. You can cancel out the waves by using another kind of energy. The next experiment will show you how.

Be a Time-Traveling Scientist
SEE HOW YOU CAN CANCEL OUT ENERGY WAVES

WHAT YOU'LL NEED
- ○ a very long piece of rope
- ○ a large room or some space outside
- ○ a friend

WHAT TO DO
1. Hold the piece of rope in your hand and ask your friend to hold the other end. Move your hand up and down quickly to make waves run down the rope to the other end. Ask your friend to keep his or her hand still. What happens?
2. Now make the waves in the rope again, but ask your friend to make waves at the same time. Try to make all the waves the same size.

WHAT HAPPENS?
When you are sending down waves, your friend will find it very hard to keep hold of the other end of the rope. It might even fly out of his or her hands! This is like what happens to the wormhole when the spaceship makes energy waves. But when waves the same size as the ones you are making are sent down the rope from the other end, your waves are canceled out, so the wormhole can stay open.

Canceling Out the Energy Waves ....

A piece of string should do the trick! Not any normal string will do, though. You have to use special cosmic string—the most powerful string in the whole universe.

Keeping a wormhole open takes incredible force. For instance, for a wormhole about 4 miles (6.5 km) across, you would need something that could press on the side of the wormhole with the pressure of a trillion boxes, each weighing a trillion tons, sitting in the palm of your hand! Nothing we know of is that strong, except perhaps something called exotic matter.

Exotic matter is the strangest stuff in the universe. It can weigh less than nothing, yet hold open a wormhole! And the best place to look for it is in pieces of cosmic string.

Cosmic string is as old as the universe. It was made in the Big Bang, when the enormous explosion forced the universe to shoot out in all directions. Different parts of the universe expanded at different rates. Some areas that expanded more slowly made long, thin tubes of space, known as cosmic string.

Be a Time-Traveling Scientist
SEE HOW COSMIC STRING
WAS MADE

WHAT YOU'LL NEED
- ○ a balloon
- ○ some glue

WHAT TO DO
1. Dot some glue very lightly onto the balloon different places and let it dry.
2. Blow up the balloon.

WHAT HAPPENS?
The dried glue prevents parts of the balloon from expanding as you blow it up. The balloon under the glue crinkles up and sometimes folds over.

If you imagine that the balloon is the universe and your blowing is the Big Bang, you can imagine how cosmic string was made.

How Long Is a Piece of Cosmic String?...

Cosmic string is longer than you could possibly imagine. In fact, pieces of cosmic string don't have ends at all. They either join together in loops or stretch right across the whole universe! They are also incredibly thin. The average piece of cosmic string is only one thousand-billion-billion-billionth of an inch wide! At the same time, it is incredibly heavy. A piece of cosmic string just a half mile long would weigh as much as the Earth.

Inside the cosmic string is space that has been trapped for billions of years. This space still contains the enormous amount of energy that was released in the Big Bang. This energy makes the space in the string very special. Called exotic matter, it has all the power you would ever need to keep your wormhole open. So, get down to the space store and get some—then you can go wormhole-hopping.

CHAPTER SEVEN

BUILD A TIME MACHINE

Congratulations! You've been chosen for a very special mission to build the world's first time-travel machine. It's a very hazardous job—and a lonely one, too—but you will be the most famous person who has ever lived! Here's your spaceship. It has the most powerful engines that have ever been made and lets you travel at half the speed of light.

Step 1

The first thing you have to do is make a wormhole here on Earth. (This is the trickiest part, because you have to figure this out for yourself. No one has discovered how to do this yet.)

Step 2

Leave one end of the wormhole attached to the Earth, put the other end in your spaceship, and blast off into space.

Step 3

Your first mission is to stay in space for about three or four years, then return to Earth. As you travel through space, time moves more slowly for you than it does on Earth. So even though three or four years may seem like a long time to you, it's not half as long as the time that will have passed on Earth. By the time you land, everyone in your class will be way ahead of you.

Step 4

Now you have to unload the end of the wormhole from your spaceship. It is connected to the wormhole you left on Earth about seventy years ago! While you relax, people start lining up to walk through the wormhole. As soon as they go through, they will find themselves on Earth at the very time you left it. Even as you watch, people from the past come through to explore the future. Congratulations! You have created a way to travel through time!

Can Time Travel Really Happen?.............

The famous scientist Stephen Hawking has said that time travel must be impossible, because if we could do it, there would be time travelers coming back to visit us all the time.

Still, time travel just might be possible. The thing about using wormholes for time travel is that you can't travel farther back in time than the moment when you made the machine. We haven't invented one yet, so it isn't surprising that there are no time travelers wandering around. However, some scientists do believe that we will one day find a way to make wormholes. And when someone does, you'll know just how to travel through it!

Other HOW TO Titles

How To...

Want to take a trip to the moon or clone your pet hamster? The *How To* books are step-by-step guides to becoming a science superstar!

How To Build a Rocket:
ISBN 0-531-14643-X

How To Split the Atom:
ISBN 0-531-14646-4

How To Build a Time Machine:
ISBN 0-531-14644-8

How To Clone a Sheep:
ISBN 0-531-14645-6

DATE DUE

DEMCO